PlutoShine

PRESS

SPACE BOOKS WITH **HEART**

PLUTO

New Horizons

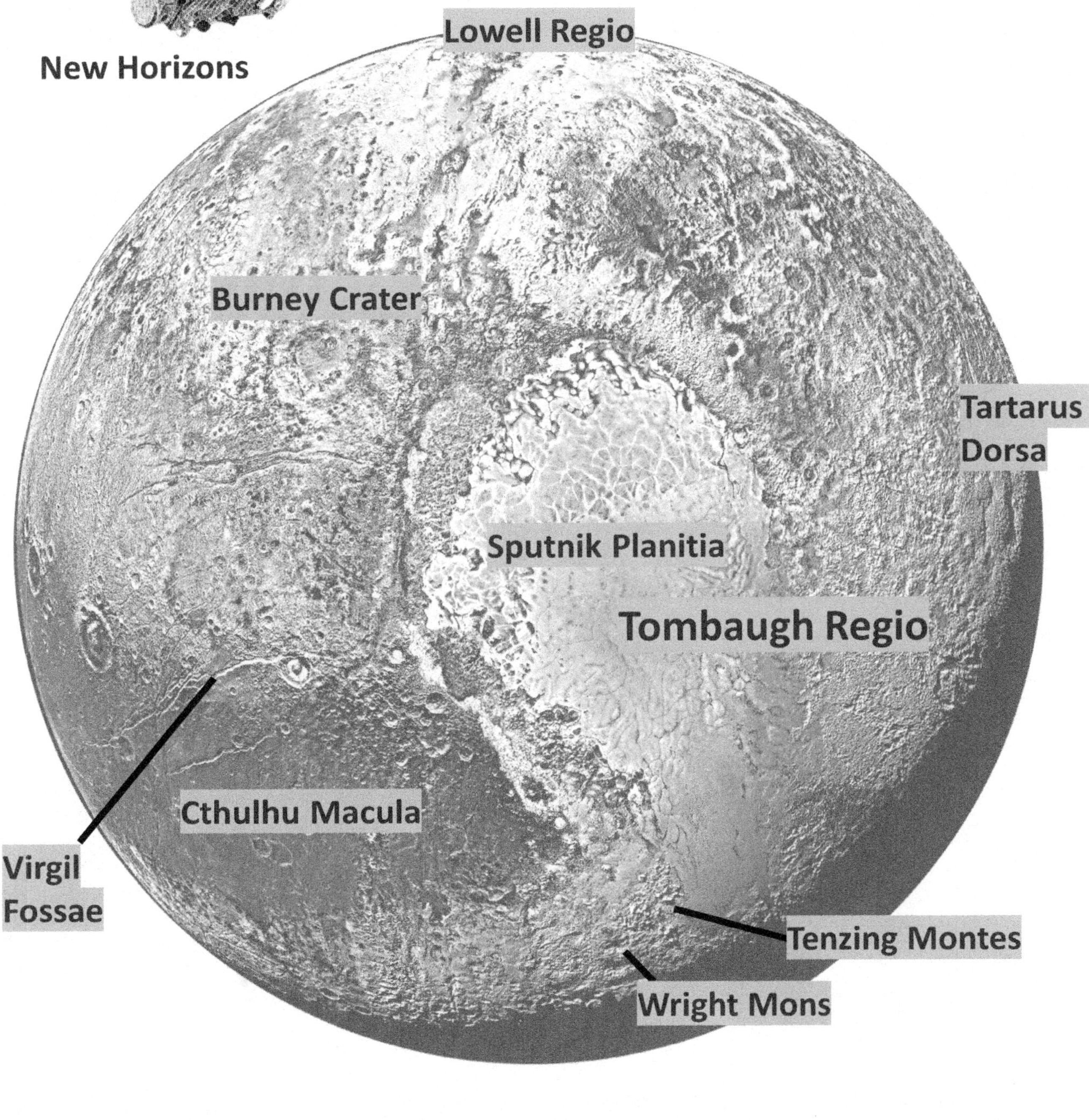
Lowell Regio
Burney Crater
Tartarus
Dorsa
Sputnik Planitia
Tombaugh Regio
Cthulhu Macula
Virgil
Fossae
Tenzing Montes
Wright Mons

Cover Image Credit: NASA/ John Hopkins University Applied Physics Laboratory/Southwest Research Institute

Image Used to Create Map: NASA/ John Hopkins University Applied Physics Laboratory/Southwest Research Institute/ZLDoyle

Pluto is composed of about
70% rock and 30% ice.

It takes about 248 years for Pluto to go around the Sun one time! So, one Pluto year is equal to 248 Earth years.

For about 20 years of its 248-year orbit, Pluto is closer to the Sun than Neptune is.

Pluto has five natural satellites. From largest to smallest they are Charon, Hydra, Nix, Kerberos, and Styx.

Pluto's moons Hydra and Kerberos may be **contact binary objects**. They appear to be composed of more than one object held together by gravity.

Pluto spins in the Kuiper belt, along with Makemake, Haumea, and hundreds of other small, round worlds, as well as countless more much smaller, irregularly shaped objects.

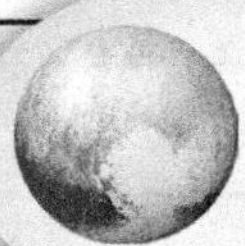

The Kuiper belt is a vast disc in the outer solar system, similar to the asteroid belt but 20 times as wide. It extends from the orbit of Neptune at 30 AU to approximately 50 AU from the Sun.

Charon is about half Pluto's diameter and about 12% of Pluto's mass. It's the largest moon relative to its host planet in the solar system!

Pluto's heart is called Tombaugh Regio (TOM-bah REDG-ee-oh). It's named after its discoverer, Clyde Tombaugh.

Pluto was discovered on February 18, 1930 by Clyde Tombaugh at Lowell Observatory in Flagstaff, Arizona.

Pluto's discoverer, Clyde Tombaugh, was only 24 years old when he discovered Pluto, and had received no college education.

Lowell Observatory hired Tombaugh to find Planet X, a hypothetical large planet thought to be hiding in the outer solar system. He found little Pluto instead!

After the discovery, many people wrote in to suggest names for Pluto. An 11-year-old British girl named Venetia Burney gets the credit for "Pluto."

Pluto's day lasts 153.3 hours. This is the length of time it takes to rotate all the way around one time.

Some scientists consider Pluto and Charon to be a **double planet** because Charon is so large relative to Pluto.

The four little moons — Hydra, Nix, Styx, and Kerberos — orbit both Pluto and Charon.

NASA's New Horizons spacecraft flew by Pluto on July 14, 2015.

The average temperature on Pluto is -387°F (-232°C).

NASA's New Horizons spacecraft was launched from Cape Canaveral Space Force Station on January 19, 2006.

The Sun is very small in Pluto's sky, but it would look more than 250 times brighter than the full Moon does from Earth!

Pluto's heart is about 1,000 miles across (1,600 km) and is composed of two geologically distinct halves.

Charon is much closer to Pluto (12,200 miles/19,640 km) than the Moon is to Earth (238,855 miles/384,400 km). This means Charon appears very large in Pluto's sky.

Pluto's moons may have formed as a result of a collision between Pluto and another space object billions of years ago.

The western half of Pluto's heart is called Sputnik Planitia. It is surprisingly smooth, without a single crater across the large region.

Pluto's coloring is unusual in that areas of high color contrast exist next to one another. The coloring of most planets, such as Mars, appears more uniform.

Pluto has two suspected **cryovolcanoes** named Wright Mons and Piccard Mons. These are volcanoes built of frozen ice that erupt a slurry of molten ice.

Charon is larger than the dwarf planet Ceres. It is the twelfth-largest natural satellite in the solar system.

Pluto's heart is composed of mostly nitrogen ice, with some methane ice and small amounts of carbon monoxide ice.

Pluto is the largest dwarf planet and the largest known transneptunian object. However, dwarf planet Eris is more massive than Pluto.

Pluto has a thin but robust atmosphere that is unlike any other in our solar system. It is composed of many layers and extends far beyond Pluto's sphere.

The New Horizons spacecraft traveled nine and a half years and three billion miles to reach Pluto.

The average distance from the Sun to Pluto is 3.7 billion miles (5.9 billion km).

Pluto's **elliptical** orbit is shaped like a flattened oval, rather than a circle. This means the distance between Pluto and the Sun varies greatly and depends on where Pluto is in its orbit.

Pluto's closest point to the Sun is 29.7 AU.
(1 AU = the distance from Earth to the Sun).
Its farthest point away is 49.3 AU.

Pluto has mountains as tall as the Rockies on Earth! The Tenzing Montes rise up to 3.9 miles (6.2 km) above the surface.

Pluto and Earth have something special in common: blue skies! They are the only two objects in the solar system known to have this feature.

Pluto's heavily cratered surfaces are billions of years old. Its smooth areas are much younger and indicate recent geological activity.

The New Horizons probe got as close as 7,800 miles (12,500 km) to Pluto's surface during the flyby.

Pluto is named for the Roman god of the dead and lord of the underworld.

In Pluto's heart, water ice hills float in a "sea" of slow-moving nitrogen ice.

The famous Disney dog character was given the name Pluto shortly after the planet Pluto was discovered and named.

Astronomer James Christy discovered Pluto's largest satellite Charon on June 22, 1978 while working to refine Pluto's orbit around the Sun.

After Pluto, New Horizons' next target was the Kuiper belt object Arrokoth, which it flew by on January 1, 2019.

New Horizons' second destination, Arrokoth, is a **planetesimal,** or planetary building block left over from the early solar system.

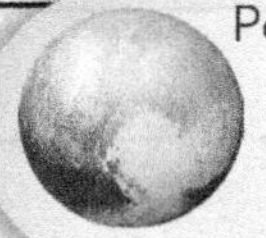

Percival Lowell was a mathematician, astronomer, and wealthy businessman who founded the observatory where Pluto was discovered and championed the search for Planet X.

The ashes of Pluto's discoverer are on board the New Horizons spacecraft, which is still cruising through the outer solar system at about 300 million miles per year.

Pluto's giant mountains are made of water ice instead of rock. At extreme low temperatures, water ice is strong like rock.

Like the planet Uranus, Pluto is tilted on its side. Its **axis** is tilted about 120°, compared to Earth's 23.5°.

New Horizons data indicates Pluto may have a few clouds at low altitude.

The surface gravity on Pluto is about 1/12 the surface gravity on Earth. As a result, if you weighed 100 pounds on Earth, you would only weigh eight pounds on Pluto!

Pluto's orbital path doesn't lie in the same plane as the Sun and the classical eight planets. It is inclined at an angle of 17°.

The tops of Pluto's mountains are dusted with methane frost.

The nitrogen ice in Pluto's heart is
softer and more plastic than water ice.
It has the ability to flow.

On Earth, nitrogen and methane take the form of gases, but on frigid Pluto, they can exist in gas *and* ice form.

In Tartarus Dorsa, a region east of Pluto's heart, lie fields of gigantic methane ice structures called **penitentes**.

Pluto's ice blades, or **penitentes**, are jagged towers of ice that rise to the height of New York skyscrapers.

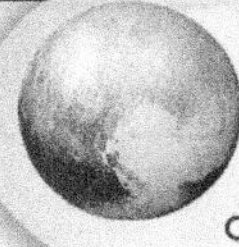

James Christy, the discoverer of Charon, named the moon after his wife Charlene. The name is also a reference to the ferryman who escorted dead souls to the underworld in Greek mythology.

The girl who named Pluto is the namesake for an instrument on board New Horizons, and for Burney Crater, a geological feature west of Pluto's heart.

Pluto is smaller than the Moon, as well as several satellites of Jupiter, Saturn, and Neptune. But it's still the 17th largest object in the entire solar system!

Pluto's moons Hydra and Nix were discovered in June 2005 by the Hubble Telescope's Pluto Companion Search Team, who were looking for hazards ahead of NASA's Pluto flyby.

Pluto's moon Hydra is named for the nine-headed serpent of Greek mythology. The letter "H" is a nod to the Hubble Telescope, while the nine heads is a nod to Pluto's status as the ninth planet.

Space objects that are sphere-shaped like Pluto have reached the state of **hydrostatic equilibrium**, which means they are massive enough to be rounded by their own gravity.

Lowell Regio, Pluto's north pole, is a region of long canyons, winding valleys, and irregularly shaped pits. It has a distinctive yellow coloring not seen elsewhere on Pluto.

Pluto's tiny moons Kerberos and Styx were discovered in 2011 and 2012 while New Horizons was en route to the Pluto System.

The vast nitrogen ice plains of Sputnik Planitia, the western half of Pluto's heart, lie over a deep basin several miles deep.

Sputnik Planitia appears to be a large crater created in a collision with another space object early in Pluto's history.

Once a gigantic hole in Pluto's crust, Sputnik Planitia filled with exotic ices over time, eventually becoming the beautiful ice plains it is today.

Pluto and Charon are **tidally locked** to each other. This means they always show the same face toward one another.

A massive crack wraps around Charon's equator. It's a series of canyons and faults four times the length of Arizona's Grand Canyon and twice as deep!

Most of Pluto's frozen surface always lies in the tropical region, thanks to the planet's extreme axial tilt.

Pluto's atmosphere is composed of nitrogen, methane, and carbon monoxide.

Pluto's atmosphere is about 100,000 times thinner than Earth's atmosphere but extends more than 13 times farther.

Every 120 years, Pluto and Charon take turns eclipsing each other once a Plutonian day (approximately every 6.4 Earth days). This goes on for about five years.

Far from being a dead world, Pluto shows many signs of activity, including a range of weather such as wind, fog, and frost.

A few worlds in our solar system have some type of snow. In Pluto's case, the snow is pieces of its atmosphere that freeze and fall to the ground.

Pluto's moon Styx is named after the river down which Charon ferried dead souls to the afterlife in Greek mythology.

A network of polygonal cells is etched in Pluto's heart. These shapes may be caused by heat rising to the surface, similar to the way oatmeal bubbles up when it's being cooked.

Pluto likely has a vast ocean of partially liquid water hidden below its surface. Fractures in key places in Pluto's crust suggest the ocean lies below its heart.

A reddish-brown, tar-like substance coats the ice of Cthulhu Macula, Pluto's dark southern region.

Pluto's snow is tinged reddish-brown as a result of sunlight interacting with methane in its atmosphere and on its surface.

Pluto's moon Nix is named for the Greek goddess of the night.

It takes eight minutes for sunlight to reach Earth. It takes five and a half hours for sunlight to reach Pluto.

The New Horizons spacecraft is roughly the size of a grand piano.

When cosmic rays interact with Pluto's methane, sticky, red-brown tholins are produced. **Tholins** are complex hydrocarbon molecules that form the basic building blocks for life.

Pluto's decades-long seasons and extreme weather contribute to the scale and diversity of landforms on its surface.

New Horizons is wrapped in a thermal, gold-colored "blanket" to protect it from the brutally cold conditions of the Kuiper belt.

In the northwest corner of Pluto's heart, windswept dunes filled with grains of methane ice lie next to soaring mountains.

Pluto's moon Kerberos is named for the three-headed dog that guards the gates to the underworld in ancient Greek myths.

Pluto's tiny moons Kerberos and Styx measure roughly six to seven miles across (10-12 km) in their longest dimension and have elongated shapes.

On August 24, 2006, a group of astronomers with the International Astronomical Union (IAU) voted to define "planet" such that it excluded Pluto and other similarly sized space objects.

According to the IAU, **planets** are gravitationally dominant and can clear other objects from their orbit, while **dwarf planets** are *not* gravitationally dominant and share space with other objects.

According to the International Astronomical Union's 2006 definitions, dwarf planets are *not* a type of planet but another class of space object altogether.

Many planetary scientists still consider Pluto and other dwarf planets to be a small type of planet. This is because dwarf planets share many similarities with rocky planets like Earth.

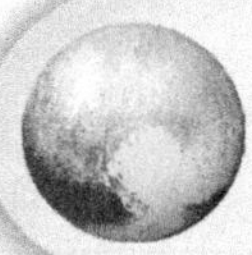

Although New Horizons was only able to capture fuzzy images of Pluto's "far side," scientists have been able to spot many interesting features such as ice blades, craters, and cliffs.

Alcyonia Lacus is a frozen nitrogen lake on Pluto. Scientists think it may have been liquid less than a million years ago.

Pluto's moons Nix and Hydra are in a chaotic and unpredictable rotation. An observer on Pluto would not see the same face of the moons from one night to the next.

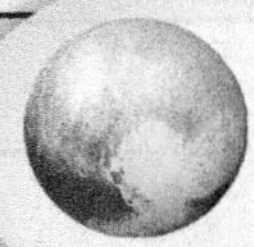

Pluto's volcanoes may have been active until relatively recently and may still be active today.

Due to its extreme tilt and orbit, Pluto's climate changes significantly over the course of about a million years.

About 800,000 years ago, Pluto was warmer and had a thicker atmosphere. It likely had liquid on its surface in the form of nitrogen lakes and rivers.

East of Pluto's heart is Sleipnir Fossa, an unusual, 360-mile-long (580 km) group of fractures that resembles an icy spider.

The western portion of Pluto's heart is relatively smooth because heat from Pluto's interior rises to the surface, churning the nitrogen ice and smoothing out craters.

Mordor Macula is a 295-mile-wide (475 km) reddish area near Charon's north pole. It may be a deposit of frozen gases captured from Pluto's escaping atmosphere.

The New Horizons probe is powered by Plutonium, a radioactive chemical element that was discovered in 1940 and named after the ninth planet.

New Horizons had to function with speed and precision. The craft had approximately nine minutes to capture the best, most centered photos of Pluto as it flew by.

Pluto's moons Nix and Hydra are about 5,000 times fainter than Pluto and about two to three times farther from Pluto than Charon.

New Horizons' trajectory took it past Jupiter, where it received a **gravity assist** from the giant planet that trimmed three years off its travel time to Pluto.

Pluto's moon Nix has a large, reddish impact crater. The red material was probably ejected from underneath Nix's water ice layer and deposited on its surface.

The largest of Pluto's little moons, Nix and Hydra are roughly 20-70 miles (32-113 km) wide.

Pluto was reclassified as a dwarf planet seven months after New Horizons launched and nine years before the spacecraft arrived.

Each of the science instruments on board New Horizons uses between two and ten watts, or about the power of a night light.

Neptune's moon Triton is only slightly larger than Pluto, and both worlds have similar surface materials, masses, and densities. This suggests Triton originated within the Kuiper belt.

Scientists think the nitrogen ice in Tombaugh Regio, Pluto's heart, may be the texture of toothpaste.

Pluto's orbit is **inclined,** or tilted, due to gravitational interaction with migrating giant planets early in the history of the solar system.

For every two orbits that Pluto makes around the Sun, Neptune makes three. This keeps Pluto's orbit stable and prevents close encounters with the giant planet next door.

Dwarf planets Haumea, Eris, and Makemake are all close in size to Pluto. When they were discovered one right after another in 2005, the debate over Pluto's planethood intensified.

When dwarf planet Eris was discovered in 2005, astronomers incorrectly estimated it was larger than Pluto. This helped spur the movement to declassify Pluto as a planet.

The total mass of all the material in the Kuiper belt is estimated to be no more than 10% of the mass of Earth.

Many Kuiper belt objects, such as Makemake and Quaoar, have a reddish coloring like that seen on Pluto's southern regions.

At high noon, Pluto is about as bright as Earth is at twilight. If you could live on Pluto, you'd have enough light during this time to read a book!

Although Pluto was discovered in 1930, astronomers did not know about the Kuiper belt, the vast third zone of our solar system, until Pluto's neighbor Albion was discovered in 1992.

When Pluto approaches its **aphelion**, or farthest point from the Sun, its atmosphere may freeze out and completely collapse, returning only as Pluto gets closer to the Sun once again.

Scientists think the ice mixture that erupts from Pluto's cryovolcanoes may have the consistency of ketchup or silly putty.

The same reddish particles, or **tholins**, that cover Pluto's surface are found in its atmosphere. They help keep Pluto's atmosphere cold, which prevents larger amounts of it from escaping into space.

Although the particles in Pluto's atmosphere are reddish, they scatter light at blue wavelengths. This is what gives Pluto its blue skies.

Pluto has a diameter of 1,477 miles (2,370 km). Dwarf planet Eris is just slightly smaller at 1,445 miles (2,325 km).

The accumulation of heavy ices in Pluto's heart-shaped glacier may be responsible for tipping Pluto on its side.

There is evidence of **tectonic activity** on Pluto. If you could live there, you would feel an occasional "plutoquake."

Because Charon is large relative to Pluto, the center of mass, or **barycenter**, between the two objects is outside Pluto. This causes Pluto and Charon to appear to orbit one another!

In the daytime when Pluto is warmer, nitrogen ice **sublimates**, or transitions directly from ice to gas without passing through a liquid state. At night, the nitrogen refreezes.

Tombaugh Regio, Pluto's heart, is in many ways the center of the planet. It drives the atmosphere and climate by circulating nitrogen winds around the planet.

Some scientists prefer the **Geophysical Planet Definition,** which puts any space object that hasn't undergone nuclear fusion and is large enough to be round under the umbrella of "planet."

Spots and streaks in Tombaugh Regio may be evidence of active geysers shooting plumes above Pluto.

Pluto has more water (in the form of ice and liquid) on its sphere than Earth does!

Pluto has a variety of landforms including flowing glaciers, canyons, valleys, channels, cliffs, scarps, pits, troughs, plains, mountains, hills, and dunes.

Scientists have detected ammonia-rich water in Pluto's Virgil Fossae and surrounding areas, suggesting water from the interior is being sprayed onto Pluto's surface through vents.

Pluto's moon Nix is only about 22 miles wide (36 km), but the reddish crater on one end is huge at 11 miles (18 km) across!

Pluto's ice volcano Wright Mons is similar in volume to Mauna Loa, one of the largest volcanoes on Earth!

The decay of radioactive elements in Pluto's core has kept it warm for billions of years. The sustained heat has helped keep Pluto's subterranean ocean partially liquid.

Active planets are changing planets. Pluto's color and brightness have changed significantly over the past few decades, and its landforms continue to change shape.

If Earth were tilted by the same amount as Pluto (120°), Texas would be in the arctic zone on Earth!

The hemisphere with Pluto's heart is always turned away from Charon.

Pluto's orbit overlaps with the orbits of other dwarf planets. However, these objects are very far away from one another, and there is no danger of collision.

The area of space per dwarf planet is more than 1000 times greater in the Kuiper belt than the area per planet in the inner solar system.

Pluto and Neptune never get closer to each other than about 16 AU, or 1.5 billion miles.

Pluto will complete its first revolution around the Sun since discovery in the year 2178.

New Horizons had to get close to Pluto and be no more than 60 miles off course. This was like hitting a golf ball from L.A. to New York and landing it in a target the size of a soup can!

Made in the USA
Monee, IL
13 September 2026

62980090R00083